Terminology

SUPERIOR:
(Cranial)
Toward the head; toward the upper part of the structure.

INFERIOR:
(Caudal)
Away from the head; toward the lower part of the structure.

ANTERIOR:
(Ventral)
Nearer to or at the front of the body.

POSTERIOR:
(Dorsal)
Nearer to or at the back of the body, or the top of the foot.

MEDIAL:
Nearer the midline of the body.

LATERAL:
Farther from the midline of the body.

PROXIMAL:
Nearer the attachment of a limb to the trunk.

DISTAL:
Away from the attachment of a limb to the trunk.

EXTERNAL:
(Superficial)
Toward the surface of the body.

INTERNAL:
(Deep)
Away from the surface of the body.

PLANTAR:
The sole or towards the sole of the foot.

Y0-CUQ-641

©1999 Bryan Edwards Publications

Movement In Joints

Abduction: Movement away from axis of trunk, as in raising arms to the side horizontally, leg sideways and scapula away from the spinal column.

Adduction: Movement toward axis of trunk, as in lowering arms to the side or leg back to position.

Circumduction: Circular movement of joint, combining movements; possible in shoulder joint, hip joint and the trunk around a fixed point.

Dorsiflexion: Movement of top of foot toward anterior tibia bone.

Eversion: Turning sole of foot outward; weight on inner edge of the foot.

Extension: Straightening: moving bones apart, as in elbow joint when hand moves away from shoulder; exception shoulder and hip joints - a return movement of the humerus or femur downward is considered extension.

Flexion: Bending: bringing bones together as in the elbow when hand is drawn to shoulder; exception shoulder and hip joints - movement of the humerus or femur to the front, upward is considered flexion.

Inversion: Turning sole of foot inward, weight on outer edge of the foot.

Plantar flexion: Movement of sole of foot downward toward the floor.

Pronation: Rotation on axis of bone, specifically applied to forearms, as in turning hand down by rotating radius on the ulna.

Internal rotation or medial rotation: Rotation with axis of bone toward body, as when humerus is turned inward.

External rotation or lateral rotation: Rotation with axis of bone away from the body, as when humerus is turned outward.

Supination: Rotation on axis of bone, specifically applied to forearm, as in turning hand up by rotating the radius on the ulna.

FLASH ANATOMY
FLASH PAKS

• •

THE JOINTS AND LIGAMENTS

• •

PUBLISHED BY

BRYAN EDWARDS PUBLICATIONS

Published by Bryan Edwards Publishing Company
Produced by Bryan E. Nash
Written and Illustrated by Flash Anatomy, Inc.
Anatomical Illustrations by Meredith Hargrave

Thanks to Randolph E. Perkins PhD.,
Assistant Professor of Physical Therapy and Anatomy
at Northwestern University Medical School, for his
contribution and advice.

© 1992 Bryan Edwards Publishing Company
All rights reserved.
Printed in the United States of America.

Printed on Recycled Paper.

Joints and Ligaments

Continued overleaf.....

Joints and Ligaments

continued....

©1999 Bryan Edwards Publications

2

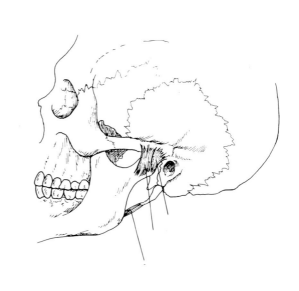

Temporomandibular Joint

Innervation: Auriculotemporal and masseteric branches of the mandibular division of the trigeminal nerve.

Arteries: Superficial temporal and maxillary arteries.

Movements: Depression, elevation, protrusion, retraction and lateral movements.

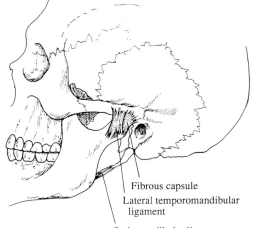

Fibrous capsule

Lateral temporomandibular ligament

Stylomandibular ligament

View: Left lateral

©1999 Bryan Edwards Publications

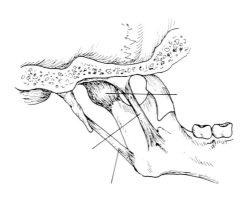

Temporomandibular Joint

Innervation: Auriculotemporal and masseteric branches of the mandibular division of the trigeminal nerve.

Arteries: Superficial temporal and maxillary arteries.

Movements: Depression, elevation, protrusion, retraction and lateral movements.

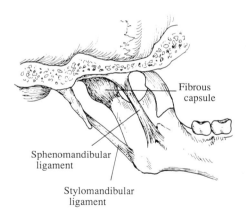

Fibrous capsule

Sphenomandibular ligament

Stylomandibular ligament

View: Left medial

©1999 Bryan Edwards Publications

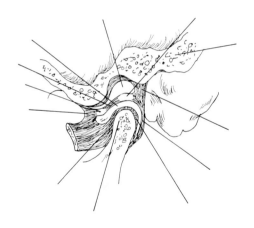

Temporomandibular Joint

Innervation: Auriculotemporal and masseteric branches of the mandibular division of the trigeminal nerve.

Arteries: Superficial temporal and maxillary arteries.

Movements: Depresion, elevation, protrusion, retraction and lateral movements.

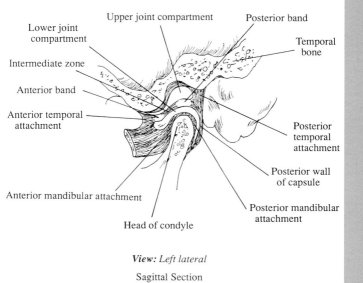

Upper joint compartment · Posterior band

Lower joint compartment

Temporal bone

Intermediate zone

Anterior band

Anterior temporal attachment

Posterior temporal attachment

Posterior wall of capsule

Anterior mandibular attachment

Head of condyle

Posterior mandibular attachment

View: Left lateral

Sagittal Section

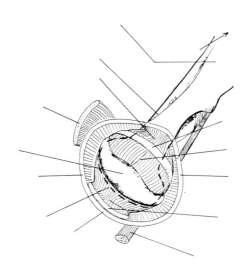

Temporomandibular Joint

Innervation: Auriculotemporal and masseteric branches of the mandibular division of the trigeminal nerve.

Arteries: Superficial temporal and maxillary arteries.

Movements: Depression, elevation, protrusion, retraction and lateral movements.

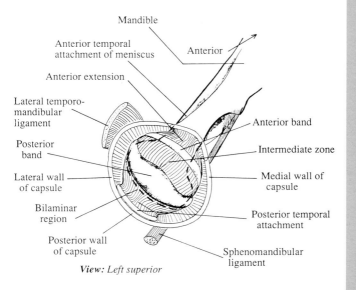

View: Left superior

©1999 Bryan Edwards Publications

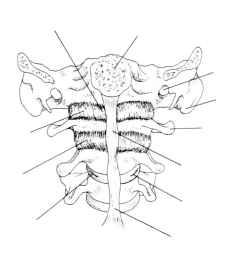

Craniovertebral Joints
Atlanto-occipital and Atlanto-axial Joints

Innvervation: Medial branches of the dorsal rami and recurrent meningeal branches of the ventral rami of adjacent spinal nerves.

Arteries: Spinal branches of the vertebral arteries.

Movements: Flexion, extension, lateral flexion, rotation and circumduction.

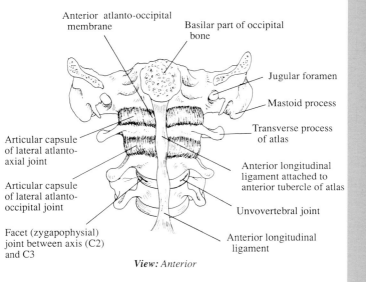

Anterior atlanto-occipital membrane

Basilar part of occipital bone

Jugular foramen

Mastoid process

Transverse process of atlas

Articular capsule of lateral atlanto-axial joint

Articular capsule of lateral atlanto-occipital joint

Anterior longitudinal ligament attached to anterior tubercle of atlas

Unvovertebral joint

Facet (zygapophysial) joint between axis (C2) and C3

Anterior longitudinal ligament

View: Anterior

©1999 Bryan Edwards Publications

Craniovertebral Joints
Atlanto-occipital and Atlanto-axial Joints

Innvervation: Medial branches of the dorsal rami and recurrent meningeal branches of the ventral rami of adjacent spinal nerves.

Arteries: Spinal branches of the vertebral arteries.

Movements: Flexion, extension, lateral flexion, rotation and circumduction.

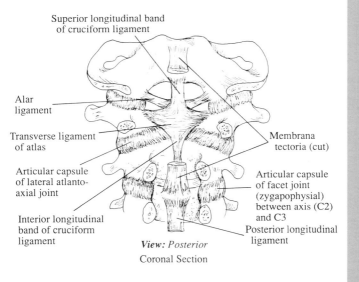

Superior longitudinal band of cruciform ligament

Alar ligament

Transverse ligament of atlas

Articular capsule of lateral atlanto-axial joint

Interior longitudinal band of cruciform ligament

Membrana tectoria (cut)

Articular capsule of facet joint (zygapophysial) between axis (C2) and C3

Posterior longitudinal ligament

View: Posterior
Coronal Section

©1999 Bryan Edwards Publications

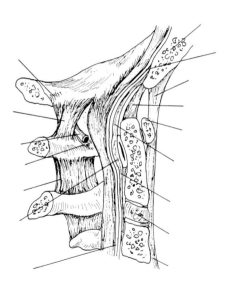

Craniovertebral Joints
Atlanto-occipital and Atlanto-axial Joints

Innervation: Medial branches of the dorsal rami and recurrent meningeal branches of the ventral rami of adjacent spinal nerves.

Arteries: Spinal branches of the vertebral arteries.

Movements: Flexion, extension, lateral flexion, rotation and circumduction.

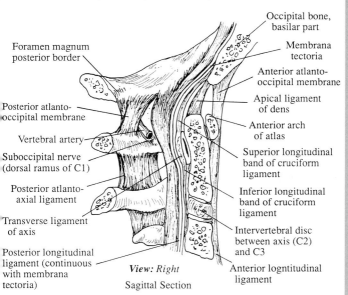

Occipital bone, basilar part

Membrana tectoria

Anterior atlanto-occipital membrane

Apical ligament of dens

Anterior arch of atlas

Superior longitudinal band of cruciform ligament

Inferior longitudinal band of cruciform ligament

Intervertebral disc between axis (C2) and C3

Anterior logntitudinal ligament

Foramen magnum posterior border

Posterior atlanto-occipital membrane

Vertebral artery

Suboccipital nerve (dorsal ramus of C1)

Posterior atlanto-axial ligament

Transverse ligament of axis

Posterior longitudinal ligament (continuous with membrana tectoria)

View: Right
Sagittal Section

©1999 Bryan Edwards Publications

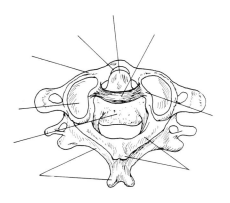

Median Atlanto-Axial Joint

Innervation: Medial branches of the dorsal rami and recurrent meningeal branches of the ventral rami of adjacent spinal nerves.

Arteries: Spinal branches of the vertebral arteries.

Movements: Flexion, extension, lateral flexion, rotation and circumduction.

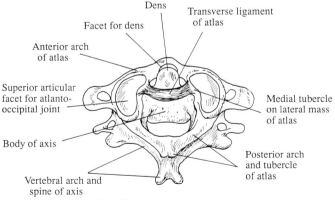

Dens

Transverse ligament of atlas

Facet for dens

Anterior arch of atlas

Superior articular facet for atlanto-occipital joint

Medial tubercle on lateral mass of atlas

Body of axis

Posterior arch and tubercle of atlas

Vertebral arch and spine of axis

View: Superior, posterior

Note: There are 3 atlanto-axial joints: 2 lateral and one median. See also cards 116 and 117.

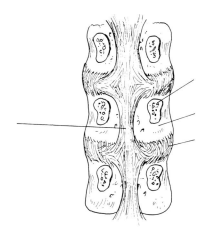

Joints of Vertebral Bodies

Innervation: Medial branches of the dorsal rami and recurrent meningeal branches of the ventral rami of adjacent spinal nerves.

Arteries: Spinal branches of the vertebral arteries.

Movements: Flexion, extension, lateral flexion, rotation and circumduction.

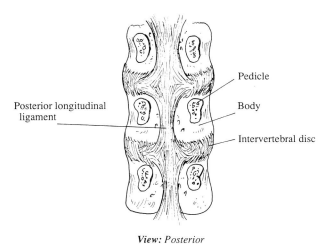

Pedicle

Body

Posterior longitudinal ligament

Intervertebral disc

View: Posterior
Coronal Section

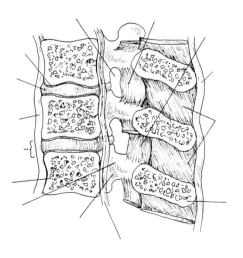

Joints of Vertebral Bodies and Arches of Lumbar Vertebrae

Innervation: Medial branches of the dorsal rami and recurrent meningeal branches of the ventral rami of adjacent spinal nerves.

Arteries: Spinal branches of the lumbar arteries.

Movements: Flexion, extension, lateral flexion, rotation and circumduction.

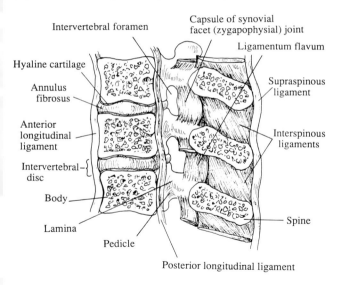

View: Right lateral

©1999 Bryan Edwards Publications

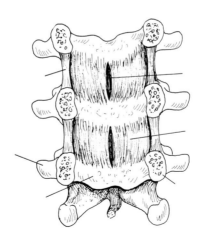

Joints of Vertebral Arches
of Lumbar Vertebrae

Innvervation: Medial branches of the dorsal rami and recurrent meningeal branches of the ventral rami of adjacent spinal nerves.

Arteries: Spinal branches of the lumbar arteries.

Movements: Flexion, extension, lateral flexion, rotation and circumduction.

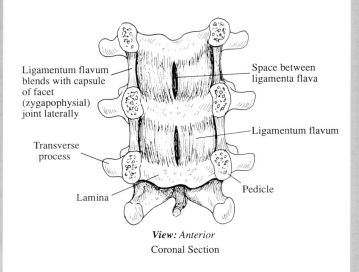

Ligamentum flavum blends with capsule of facet (zygapophysial) joint laterally

Space between ligamenta flava

Ligamentum flavum

Transverse process

Lamina

Pedicle

View: *Anterior*
Coronal Section

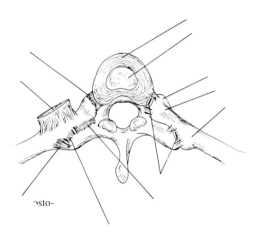

osto-

Costovertebral Joints

Innervation: Intercostal nerves.

Arteries: Spinal branches of the posterior intercostal arteries.

Movements: During inspiration slight gliding. Movements at these joints guided by the shape and direction of the articular surfaces produce evelation (pump handle movement) of the upper six ribs and eversion (bucket handle movement) of the lower six ribs.

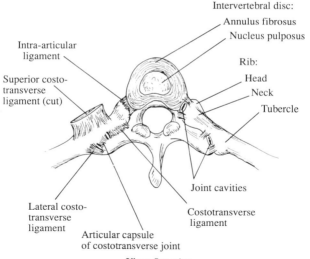

Intervertebral disc:
Annulus fibrosus
Nucleus pulposus

Intra-articular ligament

Rib:
Head
Neck
Tubercle

Superior costo-transverse ligament (cut)

Joint cavities

Costotransverse ligament

Lateral costo-transverse ligament

Articular capsule of costotransverse joint

View: Superior

©1999 Bryan Edwards Publications

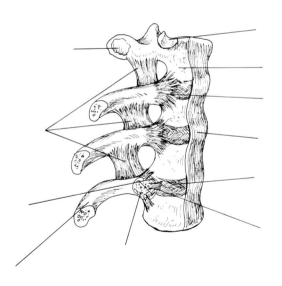

Costovertebral Joints

Innervation: Intercostal nerves.

Arteries: Spinal branches of the posterior intercostal arteries.

Movements: During inspiration slight gliding. Movements at these joints guided by the shape and direction of the articular surfaces produce evelation (pump handle movement) of the upper six ribs and eversion (bucket handle movement) of the lower six ribs.

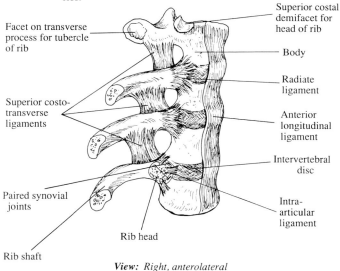

Facet on transverse process for tubercle of rib

Superior costo-transverse ligaments

Paired synovial joints

Rib shaft

Rib head

Superior costal demifacet for head of rib

Body

Radiate ligament

Anterior longitudinal ligament

Intervertebral disc

Intra-articular ligament

View: Right, anterolateral

©1999 Bryan Edwards Publications

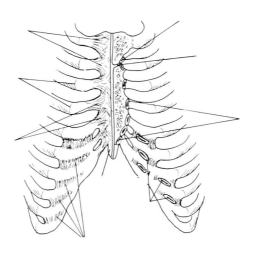

Sternocostal Joints

Innervation: Anterior cutaneous branches of the intercostal nerves.

Arteries: Branches from the internal thoracic artery.

Movements: Slight gliding sufficient for respiration.

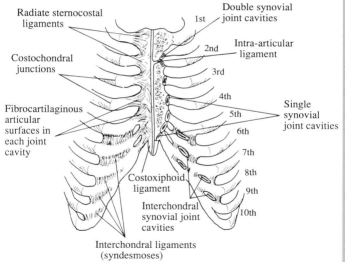

View: Anterior

Note: The synarthrosis joint of the first rib and the synovial cavities of the second to seventh are exposed by a coronal section of the sternum and costal cartilages.

©1999 Bryan Edwards Publications

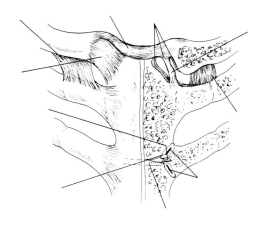

Sternoclavicular Joints

Innervation: Anterior supraclavicular and the nerve to the subclavius.

Arteries: Branches from the internal thoracic and suprascapular arteries.

Movements: (As associated with those of the scapula.) Elevation, depression, protraction, retraction, upward and downward rotation.

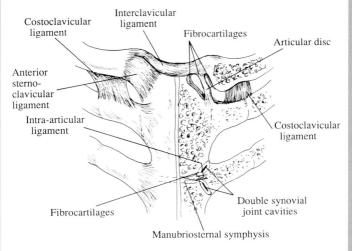

View: Anterior

Note: Right joint is intact and left is in a coronal section.

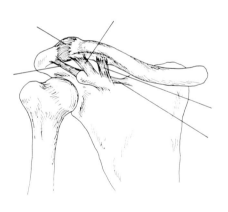

Acromioclavicular Joints

Innervation: Suprascapular and lateral pectoral nerves.

Arteries: Surpascapular and thoraco-acrominal arteries.

Movements: (As associated with those of the scapula.) Elevation, depression, protraction, retraction, upward and downward rotation.

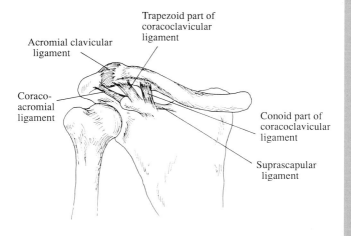

Acromial clavicular ligament

Trapezoid part of coracoclavicular ligament

Coraco-acromial ligament

Conoid part of coracoclavicular ligament

Suprascapular ligament

View: *Anterior inferior*

©1999 Bryan Edwards Publications

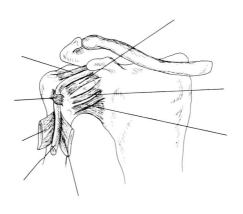

Glenohumeral Joint
(Shoulder Joint)

Innvervation: Posterior cord of the brachial plexus and the suprascapular, axillar lateral pectoral nerves.

Arteries: Anterior and posterior circumflex humeral and suprascapular arteries.

Movements: Flexion - extension, abduction - adduction, circumduction, medial and lateral rotation.

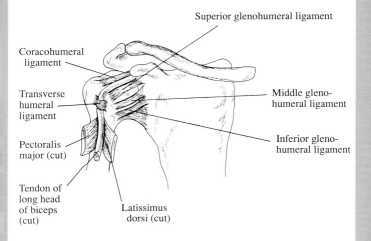

Superior glenohumeral ligament

Coracohumeral ligament

Transverse humeral ligament

Middle gleno-humeral ligament

Pectoralis major (cut)

Inferior gleno-humeral ligament

Tendon of long head of biceps (cut)

Latissimus dorsi (cut)

View: Anterior inferior

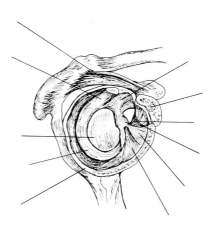

Glenohumeral Joint
(Shoulder Joint)

Innervation: Posterior cord of the brachial plexus and the suprascapular, axillar lateral pectoral nerves.

Arteries: Anterior and posterior circumflex humeral and suprascapular arteries.

Movements: Flexion - extension, abduction - adduction, circumduction, medial and lateral rotation.

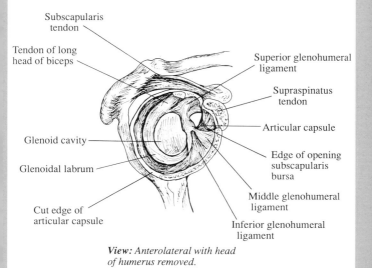

Subscapularis tendon

Tendon of long head of biceps

Superior glenohumeral ligament

Supraspinatus tendon

Articular capsule

Glenoid cavity

Edge of opening subscapularis bursa

Glenoidal labrum

Middle glenohumeral ligament

Cut edge of articular capsule

Inferior glenohumeral ligament

View: *Anterolateral with head of humerus removed.*

©1999 Bryan Edwards Publications

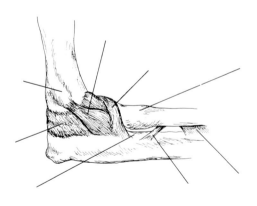

Elbow and Proximal Radio-Ulnar Joints

Innvervation: Mostly musculocutaneous and radial nerves with contributions from ulnar, median and sometimes anterior interosseous nerves.

Arteries: Anastomotic network around the elbow formed by branches of profunda brachii, brachial, radial and ulnar arteries.

Movements: Flexion and extension at the elbow joint. Supination and pronation at the proximal radio-ulnar joint.

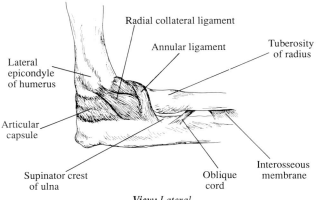

Radial collateral ligament

Annular ligament

Tuberosity of radius

Lateral epicondyle of humerus

Articular capsule

Supinator crest of ulna

Oblique cord

Interosseous membrane

View: *Lateral*

Note: The forearm is flexed at the elbow 90 degrees and is completely supinated.

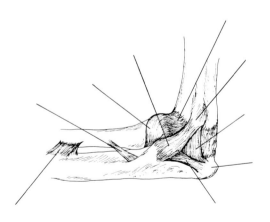

Elbow and Proximal Radio-Ulnar Joints

Innvervation: Mostly musculocutaneous and radial nerves with contributions from ulnar, median and sometimes anterior interosseous nerves.

Arteries: Anastomotic network around the elbow formed by branches of profunda brachii, brachial, radial and ulnar arteries.

Movements: Flexion and extension at the elbow joint. Supination and pronation at the proximal radio-ulnar joint.

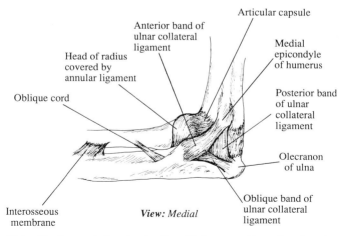

Articular capsule

Anterior band of ulnar collateral ligament

Medial epicondyle of humerus

Head of radius covered by annular ligament

Posterior band of ulnar collateral ligament

Oblique cord

Olecranon of ulna

Interosseous membrane

View: Medial

Oblique band of ulnar collateral ligament

Note: The forearm is flexed at the elbow 90 degrees and is semi-supinated.

©1999 Bryan Edwards Publications

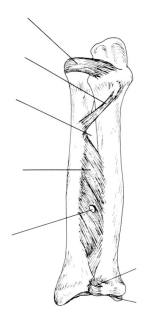

Radio-Ulnar Joints

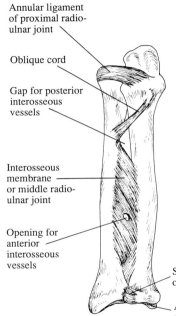

Annular ligament of proximal radio-ulnar joint

Oblique cord

Gap for posterior interosseous vessels

Interosseous membrane or middle radio-ulnar joint

Opening for anterior interosseous vessels

Sacciform recess of capsule of distal radio-ulnar joint

Articular disc of distal radio-ulnar joint

Innervation: Anterior interosseous branch of median nerve and posterior interosseous branch of radial nerve.

Arteries: Anterior interosseous atery.

Movements: Pronation and supination.

View: Anterior

©1999 Bryan Edwards Publications

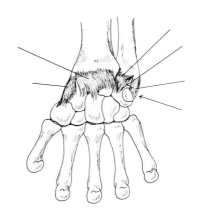

Radiocarpal Joint
(Wrist Joint)

Innervation: Anterior and posterior interosseous nerves.

Arteries: Anterior interosseous artery, anterior and posterior carpal branches of radial and ulnar arteries, palmar and dorsal metacarpal arteries and recurrent branches of deep palmar arch.

Movement: Flexion, exentsion, ulnar and radial deviation and circumduction.

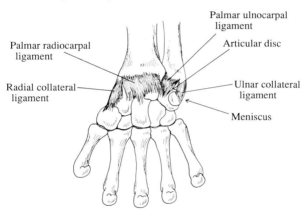

View: Palmer

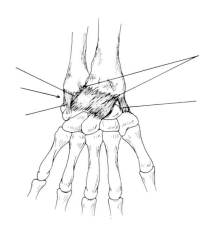

Radiocarpal Joint
(Wrist Joint)

Innervation: Anterior and posterior interosseous nerves.

Arteries: Anterior interosseous artery, anterior and posterior carpal branches of radial and ulnar arteries, palmar and dorsal metacarpal arteries and recurrent branches of deep palmar arch.

Movement: Flexion, exentsion, ulnar and radial deviation and circumduction.

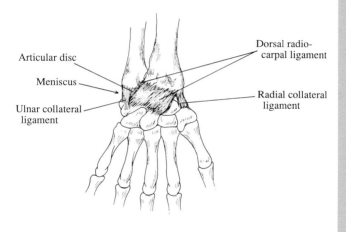

Articular disc

Meniscus

Ulnar collateral ligament

Dorsal radio-carpal ligament

Radial collateral ligament

View: Dorsal

©1999 Bryan Edwards Publications

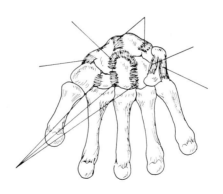

Intercarpal Joints

Innervation: Anterior and posterior interosseous nerves.

Arteries: Anterior interosseous artery, anterior and posterior carpal branches of radial and ulnar arteries, palmar and dorsal metacarpal arteries and recurrent branches of deep palmar arch.

Movement: Slight gliding at the intercarpal joints, increases the range of movemes at the wrist joint. Flexion - extension, ulnar - radial deviation and circumduction.

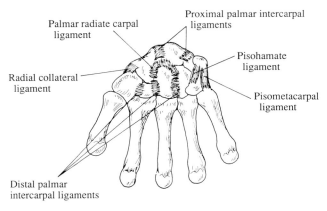

Palmar radiate carpal ligament

Proximal palmar intercarpal ligaments

Radial collateral ligament

Pisohamate ligament

Pisometacarpal ligament

Distal palmar intercarpal ligaments

View: Palmar

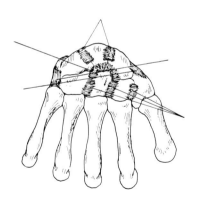

Intercarpal Joints

Innervation: Anterior and posterior interosseous nerves.

Arteries: Anterior interosseous artery, anterior and posterior carpal branches of radial and ulnar arteries, palmar and dorsal metacarpal arteries and recurrent branches of deep palmar arch.

Movement: Slight gliding at the intercarpal joints, increases the range of movemes at the wrist joint. Flexion - extension, ulnar - radial deviation and circumduction.

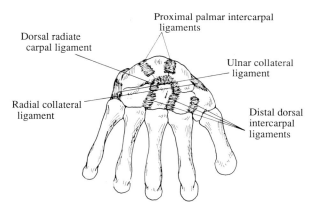

Proximal palmar intercarpal ligaments

Dorsal radiate carpal ligament

Ulnar collateral ligament

Radial collateral ligament

Distal dorsal intercarpal ligaments

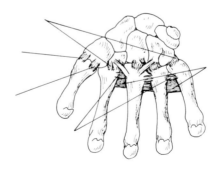

Carpometacarpal and Intercarpal Joints

Innervation: Anterior and posterior interosseous nerves.

Arteries: Anterior interosseous artery, anterior and posterior carpal branches of radial and ulnar arteries, palmar and dorsal metacarpal arteries and recurrent branches of deep palmar arch.

Movement: (For the fingers) Slight gliding which allows flexion - extension and adjunct rotation.
(For the thumb) Flexion - extension combined with conjunct medial - lateral rotation, abduction - adduction, opposition and circumduction.

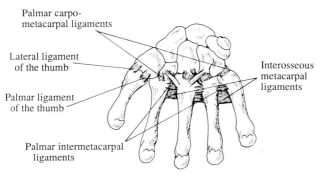

Palmar carpo-
metacarpal ligaments

Lateral ligament
of the thumb

Palmar ligament
of the thumb

Palmar intermetacarpal
ligaments

Interosseous
metacarpal
ligaments

View: Palmar

©1999 Bryan Edwards Publications

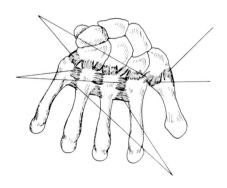

Carpometacarpal and Intercarpal Joints

Innervation: Anterior and posterior interosseous nerves.

Arteries: Anterior interosseous artery, anterior and posterior carpal branches of radial and ulnar arteries, palmar and dorsal metacarpal arteries and recurrent branches of deep palmar arch.

Movement: (For the fingers) Slight gliding which allows flexion - extension and adjunct rotation.
(For the thumb) Flexion - extension combined with conjunct medial - lateral rotation, abduction - adduction, opposition and circumduction.

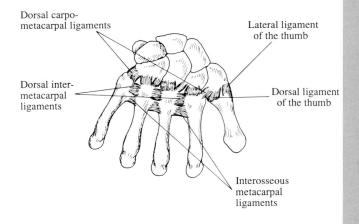

Dorsal carpo-
metacarpal ligaments

Lateral ligament
of the thumb

Dorsal inter-
metacarpal
ligaments

Dorsal ligament
of the thumb

Interosseous
metacarpal
ligaments

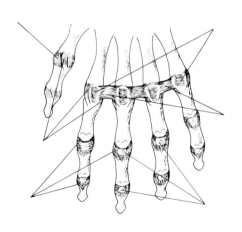

Metacarpophalangeal and Interphalangeal Joints

Innervation: Palmar digital branches of median (radial 3 1/2 digits) and ulnar (ulnar 1 1/2 digits) nerves.

Arteries: Princeps pollicis, radialis indicis, and palmar and dorsal digital arteries.

Movement: Metacarpophalangeal joints: Flexion - extension combined with conjunct rotation, abduction - adduction and circumduction.

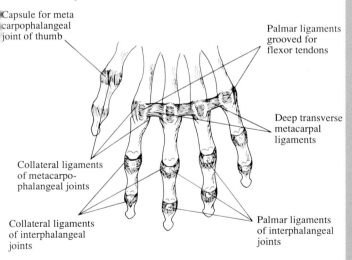

Capsule for meta carpophalangeal joint of thumb

Palmar ligaments grooved for flexor tendons

Deep transverse metacarpal ligaments

Collateral ligaments of metacarpo-phalangeal joints

Collateral ligaments of interphalangeal joints

Palmar ligaments of interphalangeal joints

View: Palmar

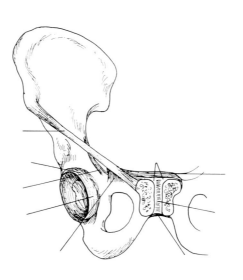

Pubic Symphysis
(Coronal Section)

Innervation: Pudendal nerve.

Arteries: Internal pudendal artery.

Movement: Slight angulation, rotation and displacement are possible. Further slight separation may occur in late gestation and during child birth.

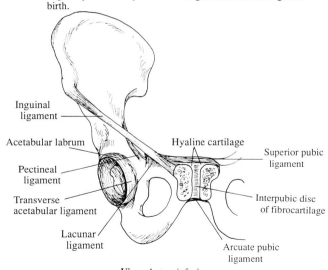

Inguinal ligament

Acetabular labrum

Pectineal ligament

Transverse acetabular ligament

Lacunar ligament

Hyaline cartilage

Superior pubic ligament

Interpubic disc of fibrocartilage

Arcuate pubic ligament

View: Anteroinferior

©1999 Bryan Edwards Publications

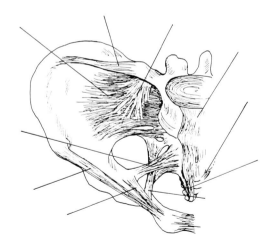

Lumbosacral, Sacrococcygeal and Sacro-iliac Joints

Innervation: Superior gluteal nerve, sacral plexus and the dorsal rami of S1 and S2 spinal nerves.

Arteries: Superior gluteal, iliolumbar and lateral sacral arteries.

Movement: For sacro-iliac joint; slight anteroposterior rotation around a transverse axis.

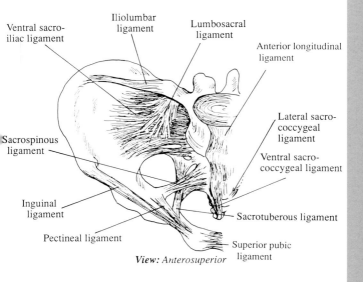

View: Anterosuperior

©1999 Bryan Edwards Publications

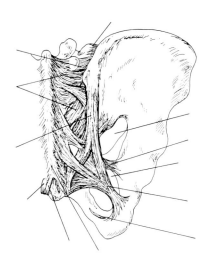

Lumbosacral, Sacrococcygeal and Sacro-iliac Joints

Innervation: Superior gluteal nerve, sacral plexus and the dorsal rami of S1 and S2 spinal nerves.

Arteries: Superior gluteal, iliolumbar and lateral sacral arteries.

Movement: For sacro-iliac joint; slight anteroposterior rotation around a transverse axis.

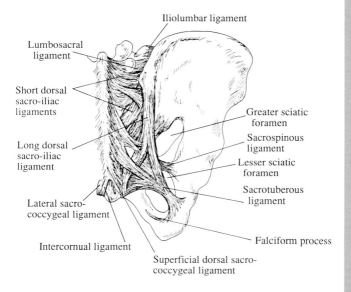

Iliolumbar ligament

Lumbosacral ligament

Short dorsal sacro-iliac ligaments

Long dorsal sacro-iliac ligament

Greater sciatic foramen

Sacrospinous ligament

Lesser sciatic foramen

Sacrotuberous ligament

Lateral sacro-coccygeal ligament

Intercornual ligament

Falciform process

Superficial dorsal sacro-coccygeal ligament

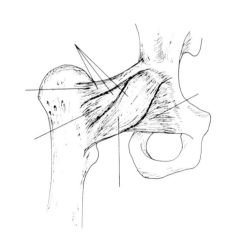

Hip (Coxal) Joint

Innervation: Femoral, obturator, accessory obturator, the nerve to quadratus femoris and the superior gluteal nerves.

Arteries: Branches from the obturator, medial circumflex femoral, and superior and inferior gluteal arteries.

Movement: Flexion - extension, abduction - adduction, circumduction, medial and lateral rotation.

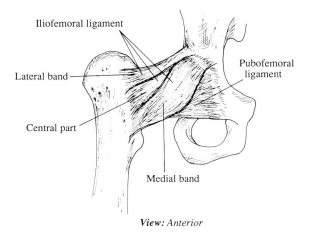

Iliofemoral ligament

Lateral band

Central part

Pubofemoral ligament

Medial band

View: *Anterior*

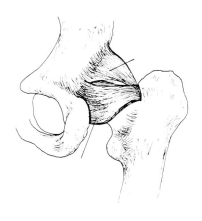

Hip (Coxal) Joint

Innervation: Femoral, obturator, accessory obturator, the nerve to quadratus femoris and the superior gluteal nerves.

Arteries: Branches from the obturator, medial circumflex femoral, and superior and inferior gluteal arteries.

Movement: Flexion - extension, abduction - adduction, circumduction, medial and lateral rotation.

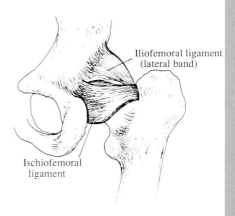

Iliofemoral ligament (lateral band)

Ischiofemoral ligament

View: Posterior

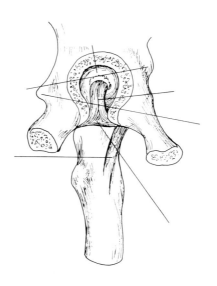

Hip (Coxal) Joint

Innervation: Femoral, obturator, accessory obturator, the nerve to quadratus femoris and the superior gluteal nerves.

Arteries: Branches from the obturator, medial circumflex femoral, and superior and inferior gluteal arteries.

Movement: Flexion - extension, abduction - adduction, circumduction, medial and lateral rotation.

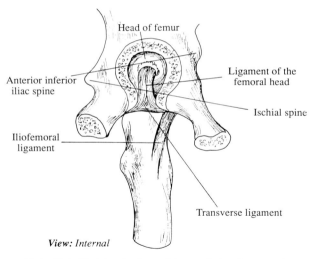

View: Internal

Note: Right hip joint; the floor of the acetabulum is removed.

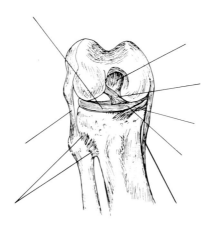

Knee Joint

Innervation: Obturator, femoral, tibial and common peroneal nerves.

Arteries: Descending genicular branches of the femoral, superior, middle, and inferior genicular branches of the popliteal, descending branch of the lateral circumflex femoral, and the anterior and posterior recurrent branches of the anterior tibial.

Movements: Flexion - extension, medial and lateral rotation.

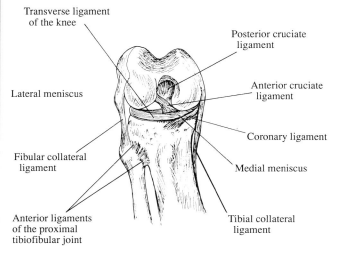

View: Anterior

Note: The right knee is in full flexion.

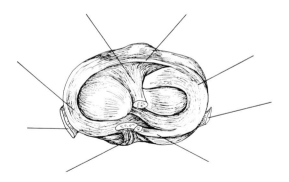

Knee Joint

Innervation: Obturator, femoral, tibial and common peroneal nerves.

Arteries: Descending genicular branches of the femoral, superior, middle, and inferior genicular branches of the popliteal, descending branch of the lateral circumflex femoral, and the anterior and posterior recurrent branches of the anterior tibial.

Movements: Flexion - extension, medial and lateral rotation.

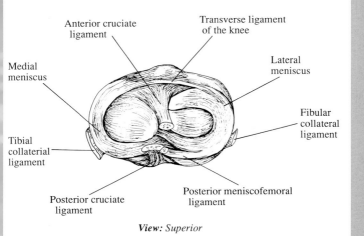

Anterior cruciate ligament

Transverse ligament of the knee

Medial meniscus

Lateral meniscus

Fibular collateral ligament

Tibial collaterial ligament

Posterior cruciate ligament

Posterior meniscofemoral ligament

View: Superior

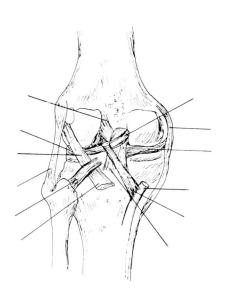

Knee Joint

Innervation: Obturator, femoral, tibial and common peroneal nerves.

Arteries: Descending genicular branches of the femoral, superior, middle, and inferior genicular branches of the popliteal, descending branch of the lateral circumflex femoral, and the anterior and posterior recurrent branches of the anterior tibial.

Movements: Flexion - extension, medial and lateral rotation.

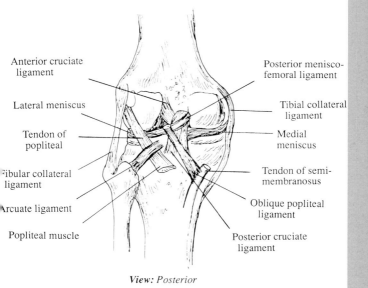

Anterior cruciate ligament

Lateral meniscus

Tendon of popliteal

Fibular collateral ligament

Arcuate ligament

Popliteal muscle

Posterior menisco-femoral ligament

Tibial collateral ligament

Medial meniscus

Tendon of semi-membranosus

Oblique popliteal ligament

Posterior cruciate ligament

View: Posterior

©1999 Bryan Edwards Publications

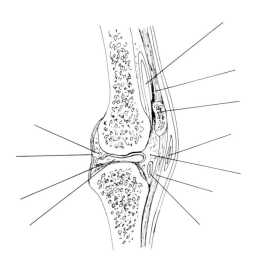

Knee Joint

Innervation: Obturator, femoral, tibial and common peroneal nerves.

Arteries: Descending genicular branches of the femoral, superior, middle, and inferior genicular branches of the popliteal, descending branch of the lateral circumflex femoral, and the anterior and posterior recurrent branches of the anterior tibial.

Movements: Flexion - extension, medial and lateral rotation.

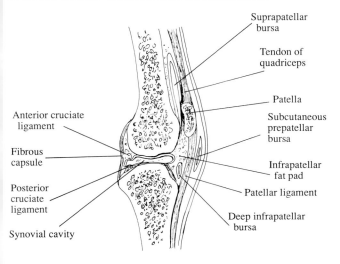

Suprapatellar bursa

Tendon of quadriceps

Patella

Subcutaneous prepatellar bursa

Infrapatellar fat pad

Patellar ligament

Deep infrapatellar bursa

Anterior cruciate ligament

Fibrous capsule

Posterior cruciate ligament

Synovial cavity

Sagittal Section

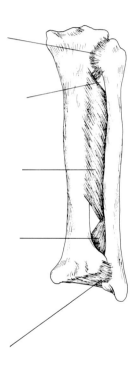

Tibiofibular Joints

Proximal Tibiofibular Joint

Innervation: Common peroneal nerve and the nerve to the popliteus, (tibial nerve).

Arteries: Anterior and posterior tibial recurrent branches of the anterior tibial artery.

Movements: Slight lateral rotation of the fibula during dorsiflexion of the ankle.

Distal Tibiofibular Joint

Innervation: Deep peroneal, tibial and saphenous nerves

Arteries: Peroneal perforating branch and medial malleolar branches of the anterior and posterior tibial arteries.

Movements: Same as proximal joint.

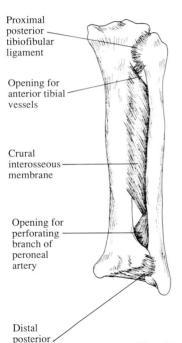

Proximal posterior tibiofibular ligament

Opening for anterior tibial vessels

Crural interosseous membrane

Opening for perforating branch of peroneal artery

Distal posterior tibiofibular ligament

View: Posterior

©1999 Bryan Edwards Publications

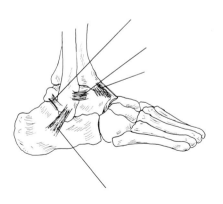

Talocrural or Ankle Joint

Innervation: Deep peroneal and tibial nerves.

Arteries: Malleolar branches of the anterior tibial and peroneal arteries.

Movements: Dorsiflexion and plantar flexion.

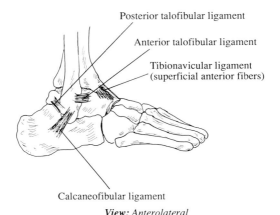

Posterior talofibular ligament

Anterior talofibular ligament

Tibionavicular ligament
(superficial anterior fibers)

Calcaneofibular ligament

View: *Anterolateral*

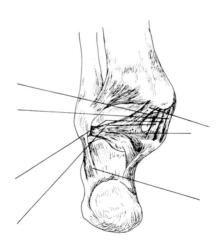

Talocrural or Ankle Joint

Innervation: Deep peroneal and tibial nerves.

Arteries: Malleolar branches of the anterior tibial and peroneal arteries.

Movements: Dorsiflexion and plantar flexion.

Parts of the deltoid ligament:

Tibiocalcaneal ligament (deep anterior fibers)

Tibionavicular ligament (super- ficial anterior fibers)

Tibiocalcaneal ligament (inter- mediate fibers)

Tibiotalar ligament (posterior fibers)

Posterior talofibular ligament and tibial slip

Calcaneofibular ligament

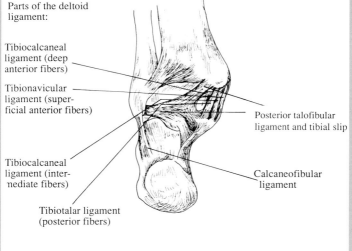

View: Posterior

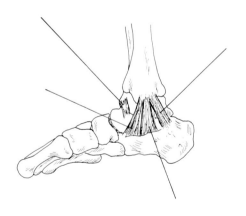

Talocrural or Ankle Joint

Innervation: Deep peroneal and tibial nerves.

Arteries: Malleolar branches of the anterior tibial and peroneal arteries.

Movements: Dorsiflexion and plantar flexion.

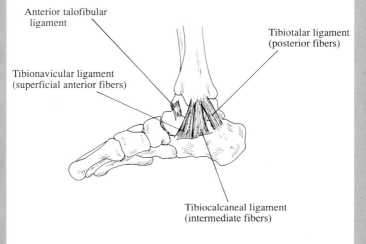

Anterior talofibular
ligament

Tibiotalar ligament
(posterior fibers)

Tibionavicular ligament
(superficial anterior fibers)

Tibiocalcaneal ligament
(intermediate fibers)

View: Anteromedial

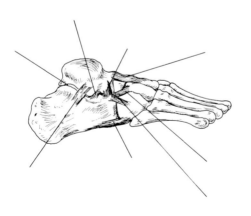

Subtalar, Intertarsal Joints
(Talocalcaneal, Talocalcaneonavicular and Calcaneocuboid Joints)

Innervation: Branches of the deep peroneal and medial and lateral plantar nerves.

Arteries: Anastomotic network around the ankle formed by branches of the anterior and posterior tibial, dorsalis pedis, peroneal, and medial and lateral plantar arteries.

Movements: Gliding and rotation at these joints produce inversion and eversion of the foot.

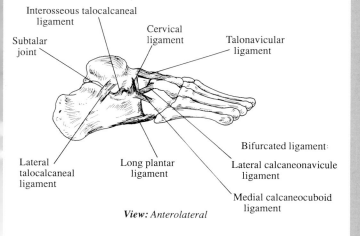

Interosseous talocalcaneal ligament

Cervical ligament

Subtalar joint

Talonavicular ligament

Lateral talocalcaneal ligament

Long plantar ligament

Bifurcated ligament

Lateral calcaneonavicule ligament

Medial calcaneocuboid ligament

View: Anterolateral

©1999 Bryan Edwards Publications

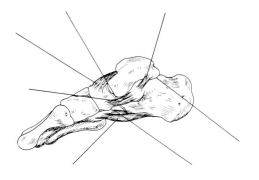

Subtalar, Intertarsal Joints
(Talocalcaneal, Talocalcaneonavicular and Calcaneocuboid Joints)

Innervation: Branches of the deep peroneal and medial and lateral plantar nerves.

Arteries: Anastomotic network around the ankle formed by branches of the anterior and posterior tibial, dorsalis pedis, peroneal, and medial and lateral plantar arteries.

Movements: Gliding and rotation at these joints produce inversion and eversion of the foot.

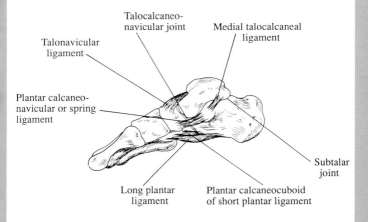

Talocalcaneo-navicular joint

Medial talocalcaneal ligament

Talonavicular ligament

Plantar calcaneo-navicular or spring ligament

Subtalar joint

Long plantar ligament

Plantar calcaneocuboid of short plantar ligament

View: *Anteromedial*

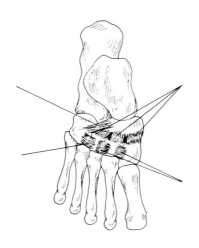

Intertarsal Joints
(Cuneonavicular, Cubideonavicular, Intercuneiform, and Cuneocubiod Joints)

Innervation: Branches of the deep peroneal and medial and lateral plantar nerves.

Arteries: Branches of the dorsalis pedis and medial and lateral plantar arteries.

Movements: Slight gliding and rotation at these joints contributes to inversion and eversion of the foot.

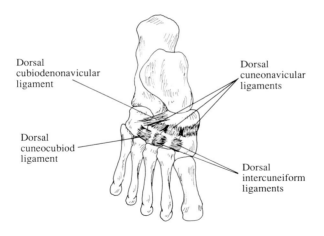

Dorsal cubiodenonavicular ligament

Dorsal cuneonavicular ligaments

Dorsal cuneocubiod ligament

Dorsal intercuneiform ligaments

View: Dorsal

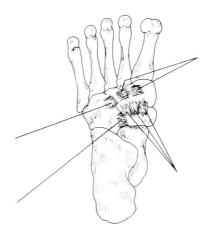

Intertarsal Joints
(Cuneonavicular, Cubideonavicular, Intercuneiform, and Cuneocubiod Joints)

Innervation: Branches of the deep peroneal and medial and lateral plantar nerves.

Arteries: Branches of the dorsalis pedis and medial and lateral plantar arteries.

Movements: Slight gliding and rotation at these joints contributes to inversion and eversion of the foot.

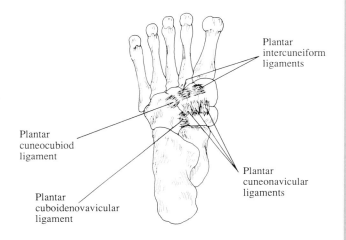

Plantar intercuneiform ligaments

Plantar cuneocubiod ligament

Plantar cuboidenovavicular ligament

Plantar cuneonavicular ligaments

View: Plantar

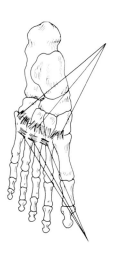

Tarsometatarsal and Intermetatarsal Joints

Innervation: Branches of the deep peroneal and medial and lateral plantar nerves.

Arteries: Plantar arch and arcuate branch of the dorsalis pedis artery.

Movements: Slight gliding during inversion and eversion of the foot.

View: Dorsal   *View: Plantar*

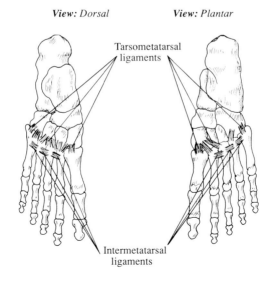

Tarsometatarsal
ligaments

Intermetatarsal
ligaments

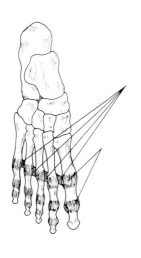
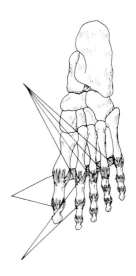

Metatarsophalangeal and Intermetatarsal Joints

Innervation: Plantar digital branches of medial plantar (medial 3 1/2 toes) and lateral plantar (lateral 1 1/2 toes) nerves.

Arteries: Plantar digital branches of the plantar arch and dorsal digital branches of arcuate branch of the dorsalis pedis artery.

Movements: Metatarsophalangeal joint: Flexion - extension, abduction - adduction. Interphalangeal joint: Flexion - extension.

View: Dorsal   *View: Plantar*

Deep transverse metatarsal ligaments

Fibrous capsules

Collateral ligaments

Plantar ligaments

©1999 Bryan Edwards Publications

Printed on Recycled Paper.

Address, Orders,
or
Editorial correspondence

Bryan Edwards
1284 East Katella Street
Anaheim, California 92805

(800)222-1775
•
(714)634-0264

ISBN
1 - 878 - 576 - 09 - 7